Route 66

Cars and Stories from the Mother Road

Etienne Psaila

Route 66: Cars and Stories from the Mother Road

First Edition: **November 2024**

ISBN: 978-1-7638074-1-9

Table of Contents

1. Introduction: The Spirit of Route 66
2. The Birth of Route 66: Paving America's Main Street
3. The Golden Age of the American Road Trip
4. Classic Cars of Route 66
5. Tales from the Road: Personal Stories of Adventure
6. Roadside Americana: Signs, Diners, and Motels
7. Cars of the Counterculture: Route 66 in the 1960s and 70s
8. Legends of the Mother Road: Local Heroes and Landmarks
9. Route 66 in Popular Culture
10. Detours and Side Roads: Exploring Beyond the Main Route
11. The Decline of Route 66
12. Preserving the Legacy: Route 66 Today
13. Modern Classics: The Cars Redefining Route 66 Today
14. The Future of the Mother Road
15. Road Trip Guide: Your Journey Down Route 66
16. Epilogue: Why Route 66 Endures
17. Appendix: Resources and Further Reading
18. Photo Gallery

The Mother Road...
Est. 1926

Historic **Route**

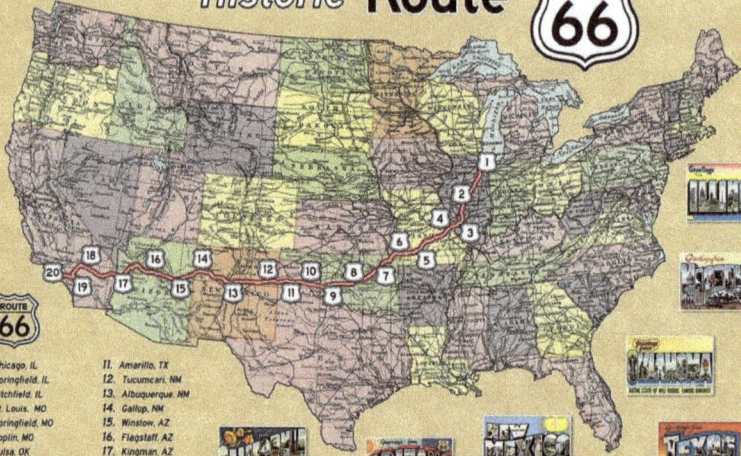

ROUTE
66

ROUTE
66

1. Chicago, IL
2. Springfield, IL
3. Litchfield, IL
4. St. Louis, MO
5. Springfield, MO
6. Joplin, MO
7. Tulsa, OK
8. Oklahoma City, OK
9. Elk City, OK
10. McLean, TX

11. Amarillo, TX
12. Tucumcari, NM
13. Albuquerque, NM
14. Gallup, NM
15. Winslow, AZ
16. Flagstaff, AZ
17. Kingman, AZ
18. Barstow, CA
19. San Bernadino, CA
20. Los Angeles, CA

Chapter 1: Introduction: The Spirit of Route 66

A Journey Like No Other

In the heart of America lies a ribbon of asphalt that transcends its humble beginnings as a highway—it is a symbol of freedom, resilience, and the ever-changing American dream. Route 66, known affectionately as the "Mother Road," is more than just a transportation route; it is an enduring testament to the spirit of adventure, the pulse of progress, and the soul of a nation. From its inception in 1926 to its designation as a Historic Route in 1985, Route 66 has captured the imagination of millions, becoming a cultural icon that reflects the story of America itself.

Spanning nearly 2,500 miles, Route 66 stretches from the bustling city streets of Chicago to the sun-drenched piers of Santa Monica. Along its course, it weaves through eight states—Illinois, Missouri, Kansas, Oklahoma, Texas, New Mexico, Arizona, and California—each contributing unique landscapes, cultures, and stories. It is a journey through the heartland of America, where travelers have discovered breathtaking natural wonders, quirky roadside attractions, and vibrant communities.

A Highway Born of Progress

In the early 20th century, America's road network was a patchwork of disconnected and poorly maintained dirt paths, making cross-country travel arduous and often dangerous. The Federal Highway Act of 1921 laid the groundwork for a national highway system, and by 1926, Route 66 was officially born. Visionary businessman Cyrus Avery, often called the "Father of Route 66," championed the idea of a highway that would connect rural America with the economic hubs of the East and West coasts.

Unlike other highways of the era, Route 66 was designed to traverse small towns and rural areas, providing farmers and small-business owners with direct access to larger markets. This made it not only a critical economic artery but also a lifeline for communities that had previously been isolated. Over time, Route 66 became a symbol of mobility, opportunity, and progress.

The Mother Road and the American Dream

Route 66 earned its nickname, the "Mother Road," from John Steinbeck's 1939 novel THE GRAPES OF WRATH. In the novel, Steinbeck describes the highway as a symbol of hope and refuge for Dust Bowl migrants seeking a better life in California during the Great Depression. For thousands of families, Route 66 was the road to survival, a path that carried them away from despair and toward new beginnings.

The highway's cultural significance grew in the post-war era, as Americans embraced the freedom of the open road. Cars became symbols of personal freedom, and Route 66 became the ultimate road trip destination. Families piled into station wagons, couples cruised in convertibles, and adventurers set out on motorcycles, all drawn by the promise of discovery and the allure of the unknown.

A Cultural Icon is Born

Route 66 has been immortalized in countless songs, movies, and television shows, becoming a cultural phenomenon that extends far beyond the United States. The song "(Get Your Kicks on) Route 66," written by Bobby Troup in 1946, celebrated the highway's vibrant spirit and solidified its place in the American psyche. The song's lyrics, which name iconic stops along the route, inspired countless travelers to embark on their own journeys.

The 1960s television series ROUTE 66 brought the highway into living rooms across America, depicting two young men exploring the country in a Corvette convertible. The show's theme of adventure and self-discovery resonated with a generation yearning for freedom and new experiences.

A Road of Contrasts and Continuity

What makes Route 66 truly special is its ability to unite contrasts: it is a road of past and present, of nostalgia and progress. On one stretch, you might encounter a perfectly preserved 1950s diner serving classic milkshakes and burgers, while on another, you might find an innovative art installation like Cadillac Ranch, where colorful graffiti-covered cars stand half-buried in the Texas soil.

The highway also represents the resilience of communities. When the Interstate Highway System threatened to make Route 66 obsolete in the 1950s and 60s, many towns along its path faced economic decline. However, grassroots efforts to preserve and promote the highway have brought renewed life to these communities, transforming Route 66 into a cherished piece of living history.

The Enduring Allure of Route 66

Today, Route 66 is more than a road; it is an experience, a journey into the heart of America's history and culture. It is a canvas painted with stories of triumph and tragedy, of ordinary people chasing extraordinary dreams. For those who travel its length, Route 66 offers more than just a scenic drive—it offers a connection to something timeless and universal.

Whether you're drawn by the romance of the open road, the charm of roadside attractions, or the rich history of the places and people along the way, Route 66 promises an adventure like no other. As we embark on this exploration of its cars, stories, and spirit, we invite you to join us in rediscovering the magic of the Mother Road, where every mile tells a tale worth remembering.

Chapter 2: The Birth of Route 66: Paving America's Main Street

A Visionary Road for a Nation on the Move

In the early 20th century, America was a country on the cusp of transformation. The automobile had become more than a luxury for the wealthy—it was emerging as a necessity for everyday Americans. Cities were booming, rural areas were seeking connection, and industries were expanding. Yet, the nation's roadways were woefully inadequate for the demands of the automobile age. It was in this context that Route 66, the "Main Street of America," was conceived.

The year was 1926, and the federal government was working to create a unified highway system. Amid this ambitious plan stood a forward-thinking entrepreneur and civic leader named Cyrus Avery, often called the "Father of Route 66." Avery's vision for a transcontinental highway would revolutionize transportation and connect the heartbeat of America in ways previously unimaginable.

The Problem with America's Roads

Before Route 66, America's roads were a chaotic patchwork of dirt paths, rutted trails, and poorly marked byways. The Lincoln Highway, established in 1913, was one of the first attempts to create a coast-to-coast route, but it lacked the infrastructure and maintenance needed for reliable travel. For many, especially those in rural areas, roads were often impassable during rain or snow, isolating communities and stifling economic growth.

By the 1920s, the explosion of automobile ownership—thanks in part to Henry Ford's Model T—highlighted the need for better roads. Farmers needed reliable routes to transport goods to markets, and businesses saw the potential of reaching customers beyond their immediate locale. Travelers, too, dreamed of the freedom to explore America by car, but they needed a road system that could take them there.

Cyrus Avery: The Father of Route 66

Cyrus Avery, an Oklahoma businessman and civic leader, recognized this need and seized the opportunity to help shape the future of American travel. Avery envisioned a highway that would link rural America to major urban centers, opening up economic opportunities for small towns and creating a vital artery for trade and tourism. His background in real estate and transportation gave him a unique understanding of how roads could transform communities.

When the U.S. Highway System was being planned, Avery championed the idea of a route that would traverse the heart of America, from Chicago to Los Angeles. Unlike other proposed highways, which tended to follow straight lines, Avery's proposed Route 66 would zigzag through smaller towns, bringing commerce and prosperity to areas often overlooked.

Avery's persuasive advocacy earned him a place on the federal committee tasked with numbering the highways. He originally suggested the number "60" for the new route, believing it conveyed importance. However, political negotiations led to the final designation: Route 66. The number had a certain rhythm and appeal, one that would later contribute to its iconic status.

A Road Designed for Progress

When Route 66 was officially established on November 11, 1926, it became one of the first highways to be fully paved, a monumental achievement in an era when most roads were still gravel or dirt. The route stretched 2,448 miles, connecting Chicago to Los Angeles, and passing through Illinois, Missouri, Kansas, Oklahoma, Texas, New Mexico, Arizona, and California. It linked farmers to markets, manufacturers to consumers, and travelers to destinations, embodying the promise of mobility and progress.

What set Route 66 apart was its intentional design to pass through rural communities rather than bypass them. This decision not only supported local economies but also created a vibrant cultural and social exchange along its length. Small towns welcomed travelers with motels, diners, and service stations, laying the groundwork for the roadside Americana that would define Route 66.

The Rise of Road Travel in America

The 1920s marked a turning point in American mobility. Cars were becoming more affordable, thanks to mass production techniques pioneered by Ford. With better vehicles came the need for better roads, and the government responded with the Federal Highway Act of 1921, which allocated funds for road construction and set standards for a national highway network.

Route 66 was not just a product of this new era; it was a catalyst for its success. The highway offered a reliable, scenic route for cross-country travel, making it easier for Americans to move, explore, and dream. Families could pack up their cars and hit the road, seeking adventure or new opportunities in faraway places. Farmers and businesses used the highway to move goods, creating a boom in commerce and industry.

Route 66: A Road of Possibilities

From its inception, Route 66 captured the imagination of those who traveled it. It was a road that offered the promise of freedom and the thrill of discovery. For Dust Bowl migrants in the 1930s, it became a lifeline, a route of escape from economic despair and environmental catastrophe. For soldiers and workers during World War II, it was a conduit for the movement of troops and materials, playing a critical role in the war effort.

But above all, Route 66 was a symbol of possibility. It reflected the American spirit of innovation and determination, a belief that with hard work and vision, anything was achievable. It wasn't just a road—it was a pathway to the future.

Legacy of the Mother Road's Creation

The creation of Route 66 marked the beginning of a new era in American transportation and culture. It connected communities, inspired dreams, and fostered the spirit of exploration. Its birth in 1926 wasn't just the designation of a highway; it was the start of a journey that would shape the identity of a nation.

As we delve deeper into the stories and cars that defined Route 66, we'll see how this iconic road became more than just a path from point A to point B—it became a road into the soul of America, where the journey itself is the destination.

Chapter 3: The Golden Age of the American Road Trip

The Dawn of a New Era

By the mid-20th century, America was on the move. World War II was over, the economy was booming, and a new era of prosperity was sweeping the nation. With increased wages, affordable cars, and a growing sense of optimism, Americans began to embrace the open road like never before. The 1940s through the 1960s marked the golden age of the road trip—a time when Route 66 and other highways became pathways to freedom, adventure, and discovery.

This was a time when the car became more than a mode of transportation. It was a cultural icon, a statement of individuality, and a gateway to the vast landscapes of the United States. Families packed their station wagons, couples set off in convertibles, and adventurous souls loaded up their motorcycles, all drawn by the promise of life on the open road. Route 66 was at the heart of this movement, offering a journey that was as much about the experience as the destination.

Life on the Open Road

Road trips during this era were defined by a sense of spontaneity and adventure. Maps were consulted, but plans were often fluid, allowing travelers to explore at their own pace. The journey itself became the highlight, as Americans rediscovered the diversity and beauty of their own country.

The cars of the time played a significant role in shaping the road trip experience. The 1950s saw the rise of spacious sedans and family-friendly station wagons like the Chevrolet Nomad and Ford Country Squire. These vehicles were built for comfort and style, with chrome accents, tailfins, and roomy interiors perfect for long drives. For the adventurous, sporty convertibles like the Ford Thunderbird or the Chevrolet Corvette added an element of glamour to the road trip.

The rhythm of a road trip was punctuated by stops at gas stations, roadside diners, and motels. Families would pull over at scenic overlooks, picnic at rest areas, and snap photos of quirky roadside attractions. The open road fostered a sense of connection—with the landscape, with strangers met along the way, and with the loved ones sharing the journey.

The Rise of Roadside Americana

One of the defining features of the golden age of road trips was the explosion of roadside attractions. As more Americans took to the highways, small towns and entrepreneurs saw an opportunity to cater to travelers with unique, eye-catching stops. Route 66 became a showcase for this phenomenon, offering a kaleidoscope of diners, motels, and novelty attractions that gave each journey its own character.

- **Motels:** The rise of the motel—a combination of "motor" and "hotel"—revolutionized road trip lodging. Unlike traditional hotels, motels were designed for convenience, offering parking spots right outside the door. Family-owned motels with neon signs like the Blue Swallow in New Mexico or the Wigwam Motel in Arizona became iconic stops on

Route 66. Their quirky architecture and personalized service created memorable experiences for travelers.

- **Diners and Cafés:** No road trip was complete without a stop at a classic American diner. These establishments, often with gleaming chrome exteriors and jukeboxes inside, served hearty meals that were perfect for travelers. Pie and coffee became staples of the road trip diet, while dishes like burgers and milkshakes offered a taste of Americana. Diners like the Cozy Dog Drive-In in Illinois, known for its corn dogs, became must-visit spots.

- **Roadside Attractions:** From the world's largest ball of twine to the Cadillac Ranch in Texas, Route 66 was dotted with attractions designed to lure curious travelers. These sites were often whimsical and quirky, embodying the spirit of fun and adventure that defined the road trip era. Whether it was posing for a photo with a giant dinosaur statue or exploring a "ghost town," these attractions added an element of surprise to every journey.

Family Adventures and Memories

For many families, the road trip was the ultimate vacation. Parents and children piled into their cars, often equipped with picnic baskets, coolers, and plenty of snacks. Games like "I Spy" and "license plate bingo" helped pass the time, while the radio played the latest hits or classic road trip tunes. Children pressed their faces to the windows, watching the changing scenery, from endless cornfields to towering mountains.

These trips often became the stuff of family lore, with stories of flat tires, unexpected detours, and serendipitous discoveries passed down through generations. For parents, the road trip was an affordable way to see the country and bond with their children. For kids, it was an adventure filled with wonder and excitement.

Route 66 as the Ultimate Road Trip Highway

Among all the highways, Route 66 stood out as the quintessential road trip destination. Its eclectic mix of landscapes, cultures, and attractions offered something for everyone. Travelers could explore the vibrant streets of Chicago, marvel at the red rocks of Arizona, and end their journey at the iconic Santa Monica Pier. Along the way, they would encounter a cross-section of America, from the friendly small-town diner owner to the artist transforming an old gas station into a gallery.

Route 66 wasn't just a road; it was a canvas for the American dream. It allowed people to step outside their routines and experience the freedom of the open road. It was a place where memories were made, stories were shared, and adventures unfolded.

The End of an Era

By the late 1960s, the golden age of the road trip began to wane. The rise of the Interstate Highway System brought faster, more direct routes that bypassed many of the small towns and attractions that had defined Route 66. Air travel became more affordable, and Americans began to look to other forms of vacationing.

Yet, the spirit of the golden age lives on. Today, travelers from around the world come to Route 66 not just for its history, but to recapture the magic of a time when the journey truly was the destination. The motels, diners, and roadside attractions that defined the era remain, standing as testaments to a golden age of travel that continues to inspire.

As we explore the stories and cars of Route 66, let us remember this golden age—a time when the open road symbolized endless possibilities and the promise of adventure. It is this spirit that keeps the Mother Road alive, ensuring its legacy endures for generations to come.

Chapter 4: Classic Cars of Route 66

The Steel Steeds of the Mother Road

Route 66 was more than just a stretch of asphalt; it was a stage where America's love affair with the automobile played out. From its earliest days, the highway saw an evolving parade of iconic vehicles, each representing the aspirations and ingenuity of its time. The cars that roared, rumbled, and cruised along the Mother Road weren't merely transportation—they were symbols of freedom, status, and the thrill of the open road.

Between the 1940s and the 1970s, the classic cars of Route 66 defined the golden age of American road travel. They were works of art on wheels, designed with sweeping curves, gleaming chrome, and a sense of individuality that spoke to the optimism of the era. These vehicles became synonymous with the road trip experience, and their stories are forever intertwined with the legend of Route 66.

The Early Icons: Pioneers of Road Travel

Before the mid-20th century, cars like the Ford Model A set the stage for the road trip revolution. Affordable, reliable, and easy to repair, the Model A was one of the first cars that allowed everyday Americans to hit the road. By the late 1930s and 1940s, larger and more powerful vehicles began to emerge, ready to take on the challenges of long-distance travel.

- **Ford Model A (1927–1931):** Though slightly predating the official designation of Route 66, the Model A was instrumental in connecting rural communities with the growing highway network. Its robust design and simple mechanics made it a favorite for early travelers braving rough roads and unpredictable weather.

- **Buick Super (1940s):** With its streamlined silhouette and advanced features, the Buick Super represented a leap forward in comfort and style. It was a car for those who wanted a touch of luxury on the open road.

The Post-War Boom: Cars of the 1950s

The 1950s marked a transformative decade for both Route 66 and the automotive industry. As soldiers returned home and families grew, the demand for larger, more stylish vehicles exploded. This era gave birth to some of the most iconic cars ever built—machines that weren't just functional but were also symbols of American prosperity and innovation.

- **Chevrolet Bel Air (1950–1957):** Perhaps the quintessential car of the Route 66 era, the Bel Air was a perfect blend of elegance and practicality. Its bold chrome grille, tailfins, and two-tone paint jobs captured the imagination of drivers. Families loved its spacious interiors, while its powerful V8 engine made it a joy to drive on the open road.

- **Ford Thunderbird (1955–1957):** The Thunderbird was Ford's answer to the Corvette, offering sporty performance with a touch of refinement. It became an instant classic, embodying the freedom and excitement of Route 66 for young couples and adventurers.

- **Cadillac Eldorado (1953–1959):** Synonymous with luxury, the Eldorado was a car that turned heads wherever it went. Its smooth ride and opulent interiors made it the vehicle of choice for those who wanted to travel in style. Driving an Eldorado down Route 66 was a statement of success and sophistication.

- **Chevy Nomad (1955–1957):** The Nomad station wagon was the ideal family car of the era. With plenty of room for kids, luggage, and even a picnic basket, it was perfect for cross-country road trips. Its distinctive design and versatility made it a favorite for travelers on Route 66.

Muscle Cars and Freedom: The 1960s and 70s

The 1960s and 70s ushered in an era of high-performance muscle cars and adventurous spirit. Route 66 became a proving ground for cars that prioritized speed and power, appealing to drivers who craved excitement and individuality. This was the era when the road trip became not just a family vacation but also a personal statement of rebellion and freedom.

- **Ford Mustang (1964–1973):** The Mustang, introduced in 1964, quickly became an icon of American muscle and style. Its affordable price, sleek design, and powerful engine options made it a favorite for Route 66 travelers who wanted performance and flair.

- **Chevrolet Corvette (1953–present):** The Corvette, with its fiberglass body and sporty design, was a dream car for many road trippers. It epitomized the spirit of adventure, offering speed and sophistication for those who wanted to make the most of the Mother Road.

- **Pontiac GTO (1964–1974):** Often considered the first true muscle car, the GTO brought raw power to the Route 66 experience. Its roaring engine and aggressive styling turned heads and made every stretch of the highway a thrill.

Practicality and Innovation: Cars for Every Traveler

While muscle cars and luxury cruisers dominated the imagination, Route 66 was also traveled by more practical vehicles, catering to everyday Americans. These cars were reliable, affordable, and designed to make long-distance travel accessible to the masses.

- **Volkswagen Beetle (1950s–1970s):** The Beetle's simplicity, reliability, and quirky charm made it a popular choice for young adventurers and budget-conscious travelers. Its ability to handle rough roads and tight spaces made it surprisingly adept for Route 66.

- **Dodge Dart (1960s):** Known for its durability and efficiency, the Dart was a workhorse for families and solo travelers alike. Its compact size and affordability made it a practical choice for navigating the winding roads and small towns along the route.

The Legacy of Classic Cars on Route 66

The cars that defined Route 66 weren't just vehicles—they were characters in the stories of the people who drove them. From the family heading west in a station wagon to the young couple cruising in a convertible, these cars embodied the dreams and ambitions of an era. They were the steel-and-chrome expressions of freedom and individuality that made Route 66 more than just a highway—it became an experience.

Today, classic car enthusiasts from around the world retrace the route in restored Bel Airs, Mustangs, and Corvettes, reliving the golden age of road travel. These cars continue to captivate the imagination, serving as reminders of a time when the journey was as important as the destination.

As we move forward in our exploration of Route 66, these cars remind us that every mile traveled was part of a larger story—a story of innovation, aspiration, and the enduring spirit of the American

Chapter 5: Tales from the Road: Personal Stories of Adventure

The Heart of Route 66

Route 66 is more than a highway—it's a thread in the fabric of countless lives. Across its nearly 2,500 miles, millions of people have traveled its path, leaving behind stories of adventure, hardship, triumph, and connection. These tales, told and retold over the decades, form the soul of the Mother Road. From families escaping the Dust Bowl to modern adventurers rediscovering its charm, the road is alive with the memories of those who've traveled it.

In this chapter, we dive into the heartwarming and adventurous stories of Route 66, shared by those who've experienced its magic firsthand. These personal accounts span decades, each offering a unique glimpse into what makes this highway an enduring symbol of the American spirit.

Escaping the Dust Bowl: A Road to Survival

For many in the 1930s, Route 66 wasn't just a road—it was a lifeline. One such story comes from the Martinez family, who fled the parched farmlands of Oklahoma during the Dust Bowl. With their few possessions loaded onto a rickety Ford Model A, they embarked on a journey westward, seeking hope in the promise of California.

"We had no choice," recalled Maria Martinez, who was a young girl at the time. "The land was dead, and so were our chances of staying. Route 66 gave us a way out—a chance to start over." Along the way, the family encountered kindness from strangers: a farmer who offered food, a mechanic who fixed their car for free, and a motel owner who let them sleep in a shed when they couldn't afford a room.

For Maria and her family, Route 66 wasn't just a road to California—it was a path to resilience and renewal.

Family Adventures: The Road Trip That Became a Tradition

In the 1950s, the Peterson family of Chicago began an annual summer tradition: a road trip down Route 66 to visit the Grand Canyon. Packed into their Chevrolet Bel Air, the Petersons would make the journey with a carefully planned itinerary— and plenty of room for spontaneity.

"We would always stop at the Cozy Dog Drive-In in Springfield for hot dogs on a stick," laughed Robert Peterson, the youngest of the family. "And Dad never missed a chance to pull over for those silly roadside attractions. My favorite was the 'World's Largest Rocking Chair' in Missouri."

The trips weren't without challenges—like the time their car overheated near Amarillo, Texas. Stranded on the side of the road, the family was helped by a local farmer who brought them water and entertained them with tales of growing up along the highway. For the Petersons, these unexpected detours often became the highlights of their trips.

Years later, Robert would take his own children on the same Route 66 journey, creating new memories while honoring the tradition his parents began.

A Love Story on Route 66

Not all Route 66 tales are about family vacations or escapes from hardship. For Julie and Sam, the Mother Road was where their love story began. In the summer of 1968, Julie, a college student from Illinois, set out on a solo road trip to California in her Volkswagen Beetle. Somewhere in Arizona, she pulled into a roadside café for lunch, where Sam, a traveling photographer, was seated at the counter.

"I saw her walk in with her dusty backpack and thought, 'That's the girl I'm going to marry,'" Sam recounted. "I asked if she wanted to share a booth, and the rest is history."

The two ended up traveling the rest of Route 66 together, sharing stories, snapping photos, and falling in love against the backdrop of the open road. They were married two years later and returned to Route 66 for their honeymoon, visiting the same spots where their adventure began.

Modern Wanderers: Rediscovering the Mother Road

In recent decades, Route 66 has become a pilgrimage for adventurers from around the world, eager to experience its nostalgic charm. One such traveler is Hiroshi Tanaka, a retiree from Japan, who dreamed of driving Route 66 ever since watching the 1960s TV series ROUTE 66.

Hiroshi fulfilled his dream in 2019, renting a vintage Mustang convertible for the journey. "I felt like I was living history," he said. "The motels, the diners, the neon signs—it was like stepping into another time."

Hiroshi documented his trip with hundreds of photographs, including a serendipitous moment when he helped a stranded biker fix a flat tire near Gallup, New Mexico. "It was a reminder that Route 66 is not just about the sights—it's about the connections you make along the way."

The Quirky and Unforgettable: Roadside Encounters

No collection of Route 66 tales would be complete without stories of the unexpected. Travelers often recount encounters with eccentric characters or experiences at offbeat roadside attractions. One such story comes from a group of friends driving a vintage camper van in the 1970s.

"We stopped at a diner in Arizona run by a man named Earl," said one of the travelers, Kathy Simmons. "Earl was a former rodeo clown, and he insisted on showing us his collection of cowboy boots—hundreds of them, all nailed to the walls. He even cooked us a free meal when we stayed to listen to his stories."

For Kathy and her friends, Earl's diner became a symbol of the unique and colorful people who make Route 66 more than just a highway.

Why These Tales Matter

The stories of Route 66 travelers—whether heartwarming, adventurous, or downright quirky—are what keep the Mother Road alive. These personal accounts capture the essence of what makes Route 66 special: the freedom of the open road, the kindness of strangers, and the unforgettable experiences that come with exploring the unknown.

For every traveler who has journeyed along Route 66, there is a story to tell. And for every story told, the legend of the

Mother Road grows, ensuring that its spirit endures for generations to come.

As we continue down the highway in this book, remember these tales—they remind us that Route 66 is more than a destination; it is a shared experience that connects us all.

Chapter 6: Roadside Americana: Signs, Diners, and Motels

The Heart of the Highway

Route 66 is more than just a road; it's a stage for a uniquely American form of art and culture. Along its nearly 2,500 miles, travelers encounter an endless parade of quirky, nostalgic stops that evoke the spirit of a bygone era. Diners with checkerboard floors, motels with neon signs promising "Clean Rooms," and bizarre roadside attractions like giant statues of lumberjacks—these are the icons of roadside Americana. They represent not only the character of Route 66 but also the ingenuity and charm of small-town America.

For many, these stops weren't just places to refuel, eat, or sleep—they were destinations in their own right. They were where memories were made, stories were exchanged, and the spirit of adventure flourished.

Neon Dreams: The Signs That Lit the Way

Nothing says Route 66 like the glow of neon. In the mid-20th century, neon signs were a beacon for travelers, drawing them to motels, diners, and gas stations with their vibrant colors and creative designs. From the iconic Blue Swallow Motel in Tucumcari, New Mexico, to the towering neon cowboy of the Big Texan Steak Ranch in Amarillo, Texas, these signs became landmarks on the Mother Road.

- **The Cultural Impact of Neon:** Neon wasn't just about catching the eye; it was about creating an identity. A neon sign wasn't merely a practical advertisement—it was a statement. The swirling script of a diner sign or the blinking lights of a motel marquee conveyed personality and promise. They said, "This is a place worth stopping for."

- **Preserving the Glow:** Today, neon signs are celebrated as an art form, with preservation efforts ensuring their continued glow. Many towns along Route 66 have embraced their neon heritage, restoring old signs and even commissioning new ones to keep the tradition alive. Travelers seeking the nostalgia of the open road often find themselves stopping just to snap photos of these luminous icons.

The Diners: A Taste of Home on the Road

Diners are the culinary soul of Route 66, offering travelers hearty meals and a warm welcome. These establishments, often family-owned, became gathering places where locals and visitors alike could share stories over a slice of pie or a cup of coffee.

- **The Classic American Diner:** With their chrome exteriors, red vinyl booths, and tabletop jukeboxes, diners along Route 66 were as much about the atmosphere as the food. Favorites like burgers, milkshakes, and breakfast served all day made these stops irresistible. The menus were simple, but the experience was rich with character.

- **Famous Diners of Route 66:** The Midpoint Café in Adrian, Texas, known for its "ugly crust" pies, and Lou Mitchell's in Chicago, a Route 66 starting-point staple, are just two examples of legendary diners that have become must-visit destinations for road trippers. Each diner has its own story to tell, adding layers of history and charm to the Route 66 experience.

Motels: Homes Away From Home

The rise of the road trip brought with it the golden age of the motel. These roadside lodgings, often run by mom-and-pop owners, catered to travelers looking for an affordable, convenient place to rest after a long day on the road.

- **The Birth of the Motel:** Motels were designed with the motorist in mind, offering the ultimate convenience—park your car right outside your room. Unlike big-city hotels, motels provided a more personal, often whimsical experience. Many were themed, with rooms decorated to reflect the local culture or geography.

- **Icons of the Mother Road:** The Wigwam Motels in Arizona and California, with their teepee-shaped rooms, and the Munger Moss Motel in Missouri, with its glowing neon sign, are among the most iconic stops on Route 66. These motels didn't just provide a place to sleep—they offered an experience, a touch of novelty that travelers cherished.

Roadside Attractions: Quirky Stops Along the Way

The spirit of Route 66 is also found in its eccentric roadside attractions. These stops, often the brainchildren of creative entrepreneurs, were designed to lure travelers off the road for a moment of wonder or amusement.

- **The World's Largest... Everything:** Along Route 66, you'll find the world's largest rocking chair in Fanning, Missouri, and the world's largest ketchup bottle in Collinsville, Illinois. These oversized curiosities were marketing genius, enticing road trippers with their sheer absurdity.

- **Unique Experiences:** Stops like Cadillac Ranch in Amarillo, Texas—an art installation featuring half-buried, spray-painted Cadillacs—offered travelers something to do as well as see. Other attractions, like ghost towns and historic trading posts, provided a glimpse into the history and culture of the region.

The Enduring Appeal of Roadside Americana

What makes these signs, diners, and motels so special? Part of their charm is their ability to evoke nostalgia, transporting travelers back to a time when the journey itself was an adventure. But their true magic lies in their humanity—they reflect the creativity, hospitality, and resilience of the people who built and maintain them.

For many modern travelers, the draw of Route 66 is the chance to experience this slice of Americana. It's not just about reaching Santa Monica or Chicago; it's about savoring a milkshake at a retro diner, snapping a photo under a neon sign, or chatting with a motel owner who has been welcoming guests for decades.

Keeping the Spirit Alive

As times have changed, many of these establishments have faced challenges. The rise of interstate highways bypassed much of Route 66, leading to the decline of many businesses. Yet, preservation efforts and renewed interest in Route 66 tourism have helped to keep the spirit of roadside Americana alive.

Whether it's through restoring neon signs, preserving historic diners, or simply sharing stories of the past, the legacy of these quirky and nostalgic stops continues to thrive. For travelers today, they offer more than just a glimpse into history—they provide a connection to the timeless allure of the open road.

Conclusion: Where the Road Leads Us

The signs, diners, and motels of Route 66 are more than landmarks—they are the soul of the highway. They remind us that the journey is not just about the miles traveled but about the places where we pause, the people we meet, and the memories we create. As we travel down the Mother Road, these icons of Americana invite us to slow down, take a closer look, and appreciate the art of the road trip.

Chapter 7: Cars of the Counterculture: Route 66 in the 1960s and 70s

The Spirit of Freedom on Four Wheels

The 1960s and 70s were a time of dramatic cultural change in America. The counterculture movement challenged traditional norms, advocating for freedom, individuality, and self-expression. Route 66, already a symbol of adventure and exploration, became a natural setting for this cultural shift. The highway was no longer just a route for families on vacation—it became a canvas for self-discovery, rebellion, and connection to a growing spirit of freedom.

During this era, cars took on new roles. They weren't just transportation—they were expressions of identity, rebellion, and dreams. Muscle cars with roaring engines, colorful VW buses adorned with peace signs, and motorcycles carrying lone travelers became the vehicles of choice for those seeking to break free from conformity. Route 66 wasn't just a road; it was a journey into the counterculture ethos.

Muscle Cars: Power and Pride on the Open Road

The 1960s saw the rise of the American muscle car, a phenomenon that captured the rebellious energy of the decade. Muscle cars like the Pontiac GTO, Chevrolet Camaro, and Dodge Charger symbolized raw power and individuality, perfectly aligned with the growing desire for freedom. These cars weren't just about getting from point A to point B—they were about how you got there, with style and speed.

- **Pontiac GTO (1964–1974):** Widely regarded as the car that started the muscle car craze, the GTO was a beast on the highway. Its V8 engine and aggressive design made it a favorite for Route 66 drivers who wanted to leave an impression.

- **Chevrolet Camaro (1967–1969):** The Camaro was Chevrolet's response to the Ford Mustang, and it quickly gained a loyal following. Its sleek lines and high-performance capabilities made it a popular choice for those cruising Route 66.

- **Dodge Charger (1966–1978):** With its bold design and massive engines, the Charger became a symbol of muscle car dominance. Whether roaring through the Mojave Desert or idling outside a diner, the Charger epitomized the spirit of Route 66 in the 1960s.

For young drivers, these cars weren't just machines—they were statements. They reflected the counterculture's rejection of the ordinary, offering an escape from the mundane through the sheer thrill of speed and power.

The VW Bus: A Rolling Revolution

While muscle cars dominated the highways, the counterculture movement had another iconic vehicle: the Volkswagen Type 2, better known as the VW Bus. Dubbed the "Hippie Van," it became a symbol of peace, love, and communal living.

- **A Mobile Home for Free Spirits:**
  The VW Bus wasn't about speed or power—it was about togetherness. Painted with bright colors, peace symbols, and psychedelic patterns, these vehicles were perfect for groups of friends or families traveling Route 66. Inside, they often housed makeshift beds, stoves, and shelves, turning the bus into a mobile home for the open road.

- **Freedom on a Budget:**
  The VW Bus was affordable and easy to maintain, making it accessible to young people who didn't have the funds for flashy muscle cars. Its reliability and adaptability made it the perfect vehicle for long road trips, allowing travelers to explore Route 66 at their own pace.

- **A Symbol of Protest and Peace:**
  For the counterculture movement, the VW Bus wasn't just transportation—it was a symbol of resistance to materialism and war. On Route 66, caravans of these colorful buses were a common sight, reflecting the spirit of rebellion and the search for a more meaningful way of life.

Motorcycles: The Lone Traveler's Journey

Motorcycles, too, became an emblem of counterculture freedom. Inspired by films like EASY RIDER (1969), Route 66 attracted bikers who sought the solitude and adventure of the open road.

- **Harley-Davidson and Choppers:**
  Custom motorcycles, or "choppers," became a

symbol of individuality. Stripped-down, elongated, and often adorned with personal touches, these bikes reflected their riders' personalities. The loud rumble of a Harley-Davidson engine echoed the counterculture's defiance of conformity.

- **A Spiritual Quest on the Road:** For bikers, Route 66 wasn't just a highway—it was a path to self-discovery. The solitary nature of motorcycle travel offered a chance to connect with the road, the landscape, and their own thoughts. Many bikers saw Route 66 as a spiritual journey, an escape from societal expectations.

The Counterculture Movement on Route 66

The counterculture movement reshaped the road trip experience, turning Route 66 into a symbol of personal freedom and exploration. Travelers weren't just seeking destinations—they were seeking meaning, authenticity, and connection.

- **Music and the Road:** Music was an integral part of this era, and it accompanied every road trip. From Bob Dylan to Janis Joplin, the sounds of the 60s and 70s provided a soundtrack to the counterculture's travels. Cars with cassette players and VW Buses with portable radios became moving concert halls, amplifying the spirit of rebellion.

- **A New Kind of Stop:** Route 66 in the counterculture era wasn't just about diners and motels. Communes, music festivals, and spiritual retreats became popular stops. Travelers would gather at places like Taos, New Mexico, or Sedona, Arizona, drawn by the promise of alternative lifestyles and community.

The Legacy of the Counterculture on Route 66

The 1960s and 70s transformed Route 66, infusing it with the energy of a generation questioning the status quo. The cars of this era—whether muscle cars, VW buses, or motorcycles—weren't just vehicles; they were symbols of identity and freedom. They carried travelers not only across miles but also through journeys of self-expression and rebellion.

Today, the echoes of the counterculture movement still resonate along Route 66. Vintage muscle cars and restored VW buses continue to retrace the highway, their drivers seeking the same sense of adventure and independence that defined the era. Route 66 remains a place where the spirit of freedom thrives, reminding us that the road is not just a way to get somewhere—it's a destination in itself.

Chapter 8: Legends of the Mother Road: Local Heroes and Landmarks

The People and Places that Keep Route 66 Alive

Route 66 is more than a highway—it's a living, breathing entity defined by the people and places along its path. Over the decades, a cast of characters, from business owners to artists and preservationists, have shaped the Mother Road's identity, turning it into a cultural treasure. Their stories, and the landmarks they've created or preserved, form the heart and soul of Route 66.

From quirky motels to larger-than-life art installations, Route 66 is home to landmarks that are as legendary as the road itself. These sites, shaped by the vision and dedication of local heroes, are more than stops on a journey—they are destinations in their own right, each with a story to tell.

The Heroes of Route 66

The survival and success of Route 66 have always depended on the people who live and work along its length. These local heroes, often small business owners, artists, and historians, have dedicated their lives to ensuring the road remains vibrant and relevant.

- **Juan Delgadillo: The Keeper of Delgadillo's Snow Cap Drive-In**
  In Seligman, Arizona, Juan Delgadillo opened the Snow Cap Drive-In in 1953. Built with scrap lumber

from a nearby railroad yard, the quirky diner quickly became a Route 66 institution. Known for its playful humor—like a "sorry, we're open" sign—Delgadillo's Snow Cap offered travelers not just burgers and shakes but also a sense of fun and whimsy. Even after Juan's passing, his family continues to operate the diner, keeping his legacy alive.

- **Angel Delgadillo: The Angel of Route 66**
  Juan's brother, Angel Delgadillo, is another legend of the Mother Road. A barber in Seligman, Angel played a pivotal role in the 1987 formation of the Historic Route 66 Association of Arizona. His efforts to preserve and promote Route 66 helped spark a movement that has kept the highway's history alive. Angel's barber shop has become a must-visit landmark, where visitors can hear firsthand accounts of Route 66's heyday.

- **Lucille Hamons: The Mother of the Mother Road**
  In Hydro, Oklahoma, Lucille Hamons ran a gas station and motel that became a haven for Route 66 travelers. Known for her kindness and hospitality, Lucille offered more than fuel and a bed—she offered a warm smile and a listening ear. Her service station, now a historic landmark, stands as a tribute to the countless small business owners who supported travelers along the Mother Road.

The Landmarks of Route 66

Route 66 is dotted with landmarks that capture the spirit of the highway. These sites, created or preserved by local heroes and artists, have become iconic symbols of the Mother Road's history and culture.

- **The Wigwam Motel (Holbrook, Arizona, and Rialto, California):**
  The Wigwam Motel, with its teepee-shaped rooms, is one of the most iconic and whimsical stops on Route 66. Built in the 1930s and 40s, these motels were part of a chain designed to attract travelers with their novelty. Though most Wigwam Motels have disappeared, the remaining locations have been lovingly restored, offering a glimpse into the playful charm of mid-century roadside architecture.

- **Cadillac Ranch (Amarillo, Texas):**
  Created in 1974 by the art collective Ant Farm, Cadillac Ranch is a surreal installation featuring ten vintage Cadillacs buried nose-first in the Texas soil. Covered in layers of graffiti, the cars invite visitors to leave their mark, making it a constantly evolving piece of interactive art. Cadillac Ranch is a testament to the creativity and eccentricity that define Route 66.

- **Blue Whale of Catoosa (Catoosa, Oklahoma):**
  Originally built as an anniversary gift by Hugh Davis for his wife in the 1970s, the Blue Whale of Catoosa quickly became a beloved roadside attraction. This cheerful, giant whale, perched on the edge of a pond, captures the whimsy of Route 66 and remains a favorite stop for families.

- **Meramec Caverns (Missouri):**
  Billed as "Jesse James' Hideout," Meramec Caverns is one of the oldest and most famous attractions on Route

66. Featuring stunning limestone formations and an air of mystery, the caverns have drawn visitors since the 1930s. Its distinctive roadside billboards, painted on barns, became a marketing hallmark of the Mother Road.

Preservationists: Guardians of the Past

As Route 66 faced decline in the mid-20th century, it was the work of dedicated preservationists that saved many of its landmarks and kept its stories alive. These individuals and organizations recognized the cultural significance of the road and fought to protect it for future generations.

- **The Route 66 Associations:** State-level associations, such as the Historic Route 66 Associations of Arizona, Missouri, and Illinois, have been instrumental in preserving the road. These groups organize events, restore landmarks, and educate the public about the history of Route 66.

- **Preserving the Neon Glow:** Efforts to restore vintage neon signs have played a key role in keeping the aesthetic of Route 66 alive. Projects like the Route 66 Neon Restoration Program ensure that the glowing lights of motels and diners continue to guide travelers along the highway.

The Living History of Route 66

What makes Route 66 truly special is that it's not a static museum piece—it's a living history, shaped by the people and places along its route. From the artists who create giant sculptures to the small business owners who keep diners and motels running, the Mother Road thrives because of their passion and dedication.

Each landmark tells a story. Each hero adds a chapter. Together, they form the tapestry of Route 66—a road that isn't just traveled but experienced, remembered, and celebrated.

Conclusion: The Legends Live On

The local heroes and landmarks of Route 66 remind us that the road's magic isn't just in its miles—it's in the people who make it unforgettable. They are the stewards of its history and the creators of its legends. Whether it's the playful glow of a neon sign, the quirky charm of a roadside attraction, or the heartfelt hospitality of a business owner, these icons of the Mother Road continue to inspire travelers from around the world.

As you journey along Route 66, take a moment to honor these legends. Their stories and landmarks are what make the road truly legendary.

Chapter 9: Route 66 in Popular Culture

A Highway Beyond Asphalt

Few roads in the world have transcended their physical presence to become cultural icons, but Route 66 has done just that. Its legend has been immortalized in music, movies, television, and art, capturing the imagination of generations. The Mother Road has come to symbolize freedom, adventure, and the spirit of discovery, making it a perfect muse for creators across the decades.

From Nat King Cole's smooth jazz anthem to Pixar's animated love letter in CARS, Route 66 has been a recurring character in popular culture. These portrayals have not only reflected the highway's unique charm but have also fueled its enduring mystique, drawing travelers from across the globe.

The Song That Started It All: "(Get Your Kicks on) Route 66"

When Bobby Troup penned the lyrics to "(Get Your Kicks on) Route 66" in 1946, he likely had no idea he was creating one of the most enduring tributes to the highway. Inspired by his own journey westward along Route 66, Troup's song celebrated the cities, towns, and landscapes that made the highway special.

- **Nat King Cole's Iconic Version:** While Troup wrote the song, it was Nat King Cole's smooth, jazzy rendition that cemented it as a classic. Released the same year, Cole's version became a hit, painting Route 66 as a road not just of asphalt but of excitement and allure. With its rhythmic name-checking of cities—Chicago, St. Louis, Amarillo, and more—the song offered a vivid snapshot of the road's journey through America.

- **A Legacy of Covers:** The song's infectious melody and celebratory lyrics have inspired countless artists, from Chuck Berry to The Rolling Stones, to record their own versions. Each rendition carries the spirit of the highway, ensuring that its legend continues to resonate across generations.

Route 66 on the Small Screen: The TV Series

In 1960, the highway took center stage in the television series ROUTE 66. The show followed two young men, Tod Stiles and Buz Murdock, as they traveled across America in a Chevrolet Corvette, tackling personal and social issues along the way. While the series wasn't confined to the literal Route 66, it embraced the highway's spirit of adventure and discovery.

- **A Journey of Self-Discovery:** Each episode featured a new location and a new set of characters, highlighting the diversity of people and places along the road. The show's themes of freedom and exploration struck a chord with audiences during a time of significant cultural change in America.

- **The Chevrolet Corvette:** The Corvette itself became a cultural icon, embodying the sleek, modern ideal of mobility and independence. Its presence in the show solidified its status as a dream car for the open road.

Route 66 in Film: The Big Screen Adventures

The silver screen has also embraced Route 66, using it as a backdrop for stories of discovery, redemption, and adventure.

- **Pixar's CARS (2006):** Pixar's CARS is perhaps the most heartfelt tribute to Route 66 in modern times. The film tells the story of Lightning McQueen, a self-centered race car who discovers the value of community and tradition in the forgotten town of Radiator Springs, inspired by real Route 66 towns.

  - **Radiator Springs as a Microcosm of Route 66:**
    The town represents the spirit of the Mother Road, highlighting the struggles of communities bypassed by the Interstate Highway System. Through characters like Doc Hudson and Sally Carrera, the film celebrates the resilience of these towns and their rich history.

  - **A Love Letter to Roadside Americana:**
    CARS is packed with nods to Route 66 landmarks, from neon-lit motels to roadside diners. Its vibrant visuals and heartfelt story introduced a new generation to the charm of the highway, sparking renewed interest in Route 66 tourism.

- **Other Films Featuring Route 66:**
  Movies like EASY RIDER (1969) and THE GRAPES OF
  WRATH (1940) have used Route 66 as a backdrop,
  showcasing its role as both a symbol of rebellion and
  a lifeline for those seeking better opportunities. These
  films reinforce the highway's status as a cultural
  touchstone for the American journey.

Art and Literature: Capturing the Spirit of the Road

Beyond music and film, Route 66 has inspired countless
works of art and literature, each offering a unique
perspective on its mystique.

- **Photography and Art:**
  Photographers like Dorothea Lange captured the
  human stories of Route 66 during the Dust Bowl era,
  while modern artists have immortalized its neon signs
  and roadside attractions in vivid detail.

- **John Steinbeck's THE GRAPES OF WRATH (1939):**
  Steinbeck's Pulitzer Prize-winning novel brought
  Route 66 to the literary forefront. Dubbed the "Mother
  Road" in its pages, the highway served as a lifeline for
  the Joad family as they journeyed westward in search
  of hope during the Great Depression. Steinbeck's
  evocative prose cemented Route 66's place in the
  American consciousness.

The Cultural Impact of Route 66

What makes Route 66 so appealing to artists, musicians, and
filmmakers? It's more than just a road—it's a symbol of

possibility. For decades, Route 66 has represented freedom, adventure, and the pursuit of the American dream. Its winding path through diverse landscapes and communities offers endless opportunities for storytelling.

- **A Global Icon:** Route 66's cultural influence extends far beyond America. Its portrayal in music, film, and art has made it a global symbol of the open road. Travelers from around the world are drawn to its legend, eager to experience the magic they've seen on screen or heard in song.

Conclusion: A Road That Inspires

Route 66's role in popular culture is a testament to its enduring allure. From the jazzy rhythms of Nat King Cole to the vibrant animation of Pixar's CARS, the highway continues to inspire creators and captivate audiences. Each song, show, and film adds to the legend of the Mother Road, ensuring that its spirit remains alive in the hearts of dreamers and adventurers.

As we continue our journey along Route 66, let us celebrate the cultural legacy that has kept its story alive for nearly a century. It's not just a road—it's a symbol of creativity, freedom, and the enduring power of the American dream.

Chapter 10: Detours and Side Roads: Exploring Beyond the Main Route

The Hidden Magic Beyond the Mother Road

Route 66 is legendary for its iconic stops and storied history, but its charm isn't confined to the well-trodden path. Beyond the main route lies a world of hidden gems and lesser-known treasures waiting to be discovered. These side roads and detours offer a quieter, more intimate connection to the landscapes, towns, and people of America's heartland.

For modern travelers, taking the road less traveled can add depth to the Route 66 experience, revealing forgotten histories, unique attractions, and moments of unexpected beauty. This chapter explores some of these offbeat detours and offers suggestions for those eager to explore beyond the Mother Road.

Hidden Gems Along the Route

Many treasures lie just off Route 66, offering an alternative perspective to the highway's celebrated landmarks.

- **The Painted Desert and Petrified Forest (Arizona):** While the Petrified Forest National Park is a known stop on Route 66, exploring its lesser-traveled trails can unveil breathtaking vistas and ancient wood turned to stone. The Painted Desert, with its vibrant hues of red, pink, and orange, offers a serene and surreal detour into nature's artistry.

- **Blue Hole (Santa Rosa, New Mexico):**
  Tucked away in Santa Rosa, this natural spring-fed pool is a hidden oasis. Known for its crystal-clear water and a constant temperature of 62°F, it's a favorite spot for swimming and scuba diving—an unexpected adventure in the New Mexico desert.

- **Elmer's Bottle Tree Ranch (Oro Grande, California):**
  A whimsical forest of "trees" made from old bottles, this quirky roadside attraction is the vision of artist Elmer Long. Each tree is a piece of folk art, and wandering through the ranch feels like stepping into a dreamlike world.

- **Devil's Rope Museum (McLean, Texas):**
  While barbed wire might not sound glamorous, this museum offers a fascinating look at how it shaped the American West. The museum also features Route 66 memorabilia, making it a unique blend of history and nostalgia.

Lesser-Known Towns with Big Stories

Some of Route 66's most memorable moments come from its small towns, where history and hospitality thrive just off the beaten path.

- **Galena, Kansas:**
  This small town, one of the first along Route 66, has a rich mining history and a charm that harks back to the early days of the Mother Road. Stop by the restored Kan-O-Tex service station, which inspired the

character of Mater in Pixar's CARS, for a taste of local pride.

- **Afton,** **Oklahoma:** Known for its Route 66 Packard Museum, this quiet town offers a glimpse into automotive history with its collection of restored Packard cars. It's a hidden gem for car enthusiasts and those seeking a nostalgic connection to the past.

- **Shamrock,** **Texas:** While its U-Drop Inn is a Route 66 icon, Shamrock's Irish-inspired flair and lesser-known attractions, like its annual St. Patrick's Day festival, make it a delightful detour.

Exploring Natural Wonders

Beyond the roadside attractions and small towns, Route 66 offers access to some of America's most stunning natural landscapes.

- **Red Rock State Park (Oklahoma):** Just a short detour from Route 66, this park features dramatic red sandstone cliffs and lush trails perfect for hiking and picnicking. It's a peaceful escape from the hustle of the highway.

- **Meteor Crater (Arizona):** Located about 20 miles off Route 66, this massive impact site is both awe-inspiring and humbling. Visitors can explore the crater's rim and learn about its history in the adjacent museum.

- **Grand Falls (Arizona):** Nicknamed "Chocolate Falls" for its muddy waters, this spectacular waterfall near Flagstaff is taller than Niagara Falls. It's a seasonal wonder, best visited after heavy rains or during spring snowmelt.

Suggestions for Modern Travelers

For today's explorers, embracing detours and side roads can add an extra layer of adventure to a Route 66 journey. Here are some tips to make the most of these hidden treasures:

- **Plan, but Stay Flexible:** While it's helpful to research stops in advance, leave room for spontaneity. Some of the best experiences come from unexpected discoveries.

- **Talk to Locals:** Stop into a diner or gas station and chat with the people who know the area best. Locals often have the inside scoop on hidden gems and unique experiences.

- **Use Technology to Your Advantage:** Apps and websites like Roadtrippers and Route 66 Navigation can help identify lesser-known attractions and provide directions to offbeat destinations.

- **Pack for Exploration:** Comfortable walking shoes, water, and a sense of adventure are essential for detours that might lead you down hiking trails or into small museums.

- **Respect the Journey:** Many of these lesser-known spots are in rural areas or small communities. Show respect for the people and places you visit by following local guidelines and leaving no trace.

The Allure of the Road Less Traveled

What makes these detours so special? They're a chance to see Route 66 through a different lens, to connect with places and stories that might otherwise be overlooked. They remind us that the true magic of the Mother Road isn't just in its main attractions—it's in the hidden corners, the quiet moments, and the discoveries made along the way.

Conclusion: Expanding the Route 66 Experience

Exploring beyond the main route offers a richer, more dynamic view of what makes Route 66 unforgettable. Whether it's the quirky charm of Elmer's Bottle Tree Ranch, the serenity of the Painted Desert, or the friendly faces in a small-town diner, these detours provide a deeper connection to the road's history and culture.

For modern travelers, the detours and side roads are an invitation to slow down, look closer, and embrace the unexpected. They remind us that while Route 66 may be a highway, it's also a journey—a mosaic of experiences that together create the legend of the Mother Road.

Chapter 11: The Decline of Route 66

The Rise of a New Era

For decades, Route 66 was the beating heart of American road travel. Its neon-lit motels, bustling diners, and vibrant communities thrived as millions of travelers used it to explore the country or seek new opportunities. However, by the mid-20th century, the Mother Road faced a challenge it could not outrun: the rise of the Interstate Highway System.

As new highways promised faster, more direct routes, the once-thriving Route 66 began to fade into obscurity. The decline was not just the loss of a road—it was the erosion of a cultural icon, a disruption of livelihoods, and a transformation of how America traveled.

The Interstate Highway System: A Modern Marvel

In 1956, President Dwight D. Eisenhower signed the Federal-Aid Highway Act, ushering in the construction of the Interstate Highway System. Inspired by Germany's Autobahn network, Eisenhower envisioned a grid of modern highways that would enhance national security, improve logistics, and reduce travel times for a rapidly growing population.

While this initiative was a milestone in transportation history, it came at a cost to highways like Route 66.

- **Bypassing the Mother Road:** The Interstate System was designed for efficiency, prioritizing direct routes over scenic or community-centered pathways. Unlike Route 66, which zigzagged through towns and cities, the interstates bypassed small communities, leaving businesses and attractions along Route 66 with dwindling traffic.

- **Fragmentation of the Route:**
As interstates like I-40, I-55, I-44, and I-15 were built, they replaced long stretches of Route 66. In many areas, the original road was abandoned, reduced to fragments of pavement, or repurposed as local streets. The cohesive journey once offered by Route 66 was now disrupted.

The Human Cost: Communities Left Behind

The decline of Route 66 had a profound impact on the communities that depended on it.

- **Business Closures:**
Motels, diners, gas stations, and roadside attractions that once thrived on the steady flow of Route 66 travelers saw their customer base vanish almost overnight. Family-owned businesses, often passed down through generations, struggled to survive as traffic shifted to the interstates.

- **Economic Decline:**
Small towns that had blossomed along Route 66 faced economic hardship. With fewer visitors stopping to spend money, local economies suffered, and some towns were left as virtual ghost towns.

- **Loss** of **Identity:**
  For many communities, Route 66 was more than a road—it was their lifeline and their identity. Its decline left a void, both economically and culturally, that was difficult to fill.

The Fight to Save Route 66

As Route 66 was gradually decommissioned in the 1970s, a growing number of people recognized the road's cultural and historical significance. Efforts to preserve its legacy began, driven by grassroots movements, preservationists, and enthusiasts.

- **Decommissioning** of **Route** **66:**
  In 1985, Route 66 was officially removed from the U.S. Highway System. This marked the end of its status as a federally recognized highway, but it also sparked a renewed interest in preserving its history.

- **Designation** as a **"Historic** **Route":**
  In the years following its decommissioning, Route 66 was rebranded as a "Historic Route." States like Missouri and Arizona led the charge, establishing organizations to promote and protect the remnants of the Mother Road.

  - In 1987, the Historic Route 66 Association of Arizona was formed, spearheading efforts to restore and celebrate key landmarks.

  - Other states followed, creating Route 66 associations and marking the road with distinctive brown and white signs, ensuring its legacy lived on.

Cultural Revival and Tourism

The designation of Route 66 as a historic route sparked a resurgence of interest in the road, fueled by nostalgia, tourism, and the efforts of preservationists.

- **Restoring   Landmarks:** Iconic landmarks like the Blue Swallow Motel in Tucumcari, New Mexico, and the Wigwam Motels in Arizona and California were restored to their former glory, attracting visitors eager to experience a piece of the past.

- **Cultural   Celebrations:** Festivals, car shows, and Route 66-themed events became popular, drawing enthusiasts from around the world. These celebrations not only honored the road's history but also injected new life into the communities along its path.

- **Global   Appeal:** Route 66 became a symbol of American culture and freedom, attracting international tourists who viewed it as the ultimate road trip experience. For many, driving Route 66 was not just a journey but a pilgrimage to an iconic piece of Americana.

Preservation in the Digital Age

Today, Route 66 continues to evolve, with modern technology playing a role in its preservation and promotion.

- **Mapping and Apps:**
  Digital tools like GPS apps and Route 66 travel guides help modern travelers navigate the fragmented sections of the road, ensuring they can experience its full charm.

- **Virtual Tours:**
  Online resources and virtual tours allow people to explore Route 66's history and landmarks from anywhere in the world, keeping its story alive for new generations.

A Road Reclaimed

Though Route 66 may never again serve as the primary artery of cross-country travel, its decline has been met with resilience and determination. What was once dismissed as an obsolete highway is now celebrated as a cultural treasure. The Mother Road's story is not one of loss but of reinvention—a reminder that even in decline, a legend can endure.

Conclusion: The Spirit of the Mother Road Lives On

The decline of Route 66 marked the end of an era, but it also sparked a movement to preserve its legacy. Today, the road stands as a symbol of resilience and nostalgia, a place where travelers can connect with the past and discover the magic of the open road.

For those who journey along its path, Route 66 remains a living history—a testament to the enduring spirit of the Mother Road and the people who refused to let it fade away.

Chapter 12: Preserving the Legacy: Route 66 Today

A Road Worth Saving

Route 66, once the beating heart of America's highway system, faced near extinction during the rise of the Interstate Highway System. Yet, the Mother Road has not only survived—it has thrived, thanks to the tireless efforts of preservationists, local communities, and road trip enthusiasts. Today, Route 66 is more than a highway; it is a living museum, a cultural icon, and a testament to the enduring spirit of the open road.

This chapter explores the efforts to maintain and celebrate Route 66 and how modern road trippers have embraced its legacy, ensuring it continues to inspire new generations.

Grassroots Movements: The Guardians of Route 66

The preservation of Route 66 began with passionate individuals and small groups who recognized the road's historical and cultural value.

- **The Birth of Preservation Efforts:** After Route 66 was decommissioned in 1985, grassroots organizations like the Historic Route 66 Association of Arizona emerged to advocate for its protection. These groups worked to restore landmarks, promote tourism, and educate the public about the road's significance.

- **State and Local Support:** State Route 66 associations in Illinois, Missouri, Oklahoma, and other states have played a key role in preserving the highway. Their efforts include organizing clean-up drives, restoring vintage signage, and hosting events to celebrate the road's history.

- **National Recognition:** In 1999, the Route 66 Corridor Preservation Program was established by the National Park Service to provide grants and technical assistance for preserving Route 66 landmarks. This program has been instrumental in safeguarding the road's legacy.

Reviving Landmarks and Attractions

A major focus of preservation efforts has been the restoration of Route 66's iconic landmarks, ensuring they remain vibrant for future travelers.

- **Neon Revival:** Restoring vintage neon signs has been a hallmark of Route 66 preservation. Signs like the Blue Swallow Motel in Tucumcari, New Mexico, and the Ariston Café in Litchfield, Illinois, now shine as brightly as they did in their heyday.

- **Restored Motels and Diners:** Family-owned motels like the Wigwam Motel in Holbrook, Arizona, and historic diners like Lou Mitchell's in Chicago have been lovingly restored, offering travelers an authentic taste of the past.

- **Roadside** **Attractions:**
  Quirky attractions like the Cadillac Ranch in Texas and the Blue Whale of Catoosa in Oklahoma have been preserved and enhanced, drawing visitors eager to experience the whimsy of the Mother Road.

Modern Road Trippers: Rediscovering Route 66

The revival of Route 66 is fueled by a new wave of travelers seeking nostalgia, adventure, and connection.

- **The** **Nostalgia** **Factor:**
  For many, driving Route 66 is a chance to step back in time, reliving the golden age of the American road trip. Classic cars, retro diners, and vintage motels evoke a sense of simpler, more adventurous times.

- **Cultural** **Pilgrimage:**
  Route 66 has become a pilgrimage for travelers from around the world, eager to experience the legendary highway. International tourists, in particular, view Route 66 as a quintessential American experience, blending history, culture, and scenic beauty.

- **Community** **Engagement:**
  Modern road trippers often engage with the local communities along the route, supporting small businesses and learning about the unique histories of the towns they pass through. This interaction helps sustain the communities and keeps the spirit of Route 66 alive.

Technology and the Digital Age

The preservation and promotion of Route 66 have been bolstered by modern technology, making it easier for travelers to navigate and explore the highway.

- **Route 66 Apps and Maps:** Apps like Route 66 Navigation provide turn-by-turn directions, highlight must-see stops, and share historical tidbits, ensuring travelers can fully experience the route even as sections of it become harder to find.

- **Social Media and Online Communities:** Platforms like Instagram, Facebook, and YouTube have allowed enthusiasts to share their journeys, showcase landmarks, and connect with a global community of Route 66 fans. Hashtags like #Route66 keep the highway in the public eye, inspiring more people to explore it.

- **Virtual Tours and Digital Archives:** For those unable to visit in person, virtual tours and online archives offer a way to experience Route 66's history and charm. These resources ensure that the road's legacy reaches even those who can only dream of traveling it.

Events and Festivals: Celebrating Route 66

Annual events and festivals bring Route 66 to life, drawing enthusiasts from near and far.

- **Route 66 Festivals:** Events like the Route 66 International Festival celebrate the highway's history and culture with car shows, live music, and community gatherings.

- **Car Cruises and Rallies:** Classic car clubs often organize Route 66 cruises, where vintage cars retrace the highway's path, recreating the magic of mid-century road trips.

- **Local Celebrations:** Towns along Route 66 host their own events, from chili cook-offs to historical reenactments, showcasing their unique contributions to the Mother Road's story.

The Future of Route 66

While Route 66 faces challenges, including aging infrastructure and evolving travel trends, its future remains bright thanks to ongoing efforts to preserve and promote it.

- **Expanding Preservation Programs:** Advocates continue to push for Route 66's designation as a National Historic Trail, a move that would provide federal recognition and funding for its preservation.

- **Sustainable Tourism:** As more travelers embrace sustainable practices, Route 66's small businesses and historic landmarks stand to benefit from a growing emphasis on community-based tourism.

- **New Generations of Travelers:** With each passing year, new generations discover Route 66, drawn by its timeless appeal and the stories it has to tell. Their enthusiasm ensures that the road will remain relevant and celebrated for years to come.

Conclusion: A Legacy Worth Preserving

Route 66 is more than a highway—it's a symbol of adventure, resilience, and the enduring power of the American dream. Through the efforts of preservationists, local communities, and passionate travelers, the Mother Road continues to thrive as a living history that inspires and delights.

For modern road trippers, Route 66 offers a chance to connect with the past while creating new memories. As they journey along its winding path, they carry forward the legacy of the Mother Road, ensuring that its spirit endures for generations to come.

Chapter 13: Modern Classics: The Cars Redefining Route 66 Today

The Road Meets the Future

The rumble of V8 engines, the hum of well-tuned carburetors, and the purr of classic cars have always been synonymous with Route 66. Yet, as the road evolves, so do the cars that traverse it. Today, the Mother Road sees an eclectic mix of vehicles—from lovingly restored classics to cutting-edge electric cars—each carrying forward the legacy of exploration and freedom that Route 66 represents.

Modern travelers on Route 66 are redefining what it means to experience the open road. Some aim to relive the golden era of road trips in restored Chevys and Fords, while others embrace the future with Teslas and Rivians gliding silently through desert landscapes. This chapter explores the vehicles that bring Route 66 to life today, blending nostalgia with innovation.

Restoring the Classics: Keeping the Spirit Alive

For many, Route 66 is best experienced behind the wheel of a classic car, where every mile echoes the highway's storied past. Restorations of vintage cars have become a passionate pursuit for enthusiasts who want to relive the heyday of the Mother Road.

- **Chevrolet Bel Airs and Ford Mustangs:** Restored Chevrolet Bel Airs, with their chrome accents and two-tone paint jobs, are a common sight along Route 66, embodying the spirit of the 1950s. Similarly, the Ford Mustang, a symbol of 1960s muscle and freedom, continues to turn heads as it roars down the highway.

- **Preservation Meets Performance:** Restorations today often strike a balance between authenticity and modern functionality. While the exteriors remain faithful to their original designs, many classic cars are retrofitted with modern engines, air conditioning, and safety features, allowing drivers to enjoy the nostalgic aesthetic without sacrificing comfort or reliability.

- **The Community of Enthusiasts:** Classic car clubs and gatherings along Route 66 foster a sense of camaraderie among drivers. Events like vintage car rallies and Route 66-themed car shows celebrate these timeless machines, bringing together enthusiasts who share a love for the road and its history.

The Rise of Electric Roadsters: A Silent Revolution

While classics dominate the nostalgic landscape of Route 66, the highway is also becoming a proving ground for the cars of tomorrow. Electric vehicles (EVs), with their futuristic designs and sustainable credentials, are reshaping how travelers experience the Mother Road.

- **Teslas and Rivians:** Tesla Model 3s and Model Ys have become a common sight along Route 66, their sleek designs contrasting sharply with the road's vintage Americana aesthetic. Rivian's R1T electric trucks, with their rugged capability, are ideal for adventurers seeking to explore the side roads and natural wonders off the main route.

- **Charging the Mother Road:** The increasing availability of EV charging stations along Route 66 has made electric road trips more accessible. Historic stops like the Blue Swallow Motel in Tucumcari, New Mexico, now feature EV chargers, blending past and future seamlessly.

- **Sustainability and the Modern Road Trip:** For environmentally conscious travelers, electric vehicles offer a way to enjoy Route 66 without contributing to its environmental footprint. This silent revolution ensures that the spirit of exploration endures while adapting to modern sensibilities.

Custom Builds and Modern Interpretations

In addition to restorations and EVs, custom-built vehicles are making their mark on Route 66. These cars often fuse retro aesthetics with contemporary technology, creating unique rides that honor the past while embracing the future.

- **Restomods:** Restomods—classic cars updated with modern components—are particularly popular on Route 66. A restomod might look like a 1967 Chevy Camaro on the outside but feature a modern drivetrain, touchscreen navigation, and fuel-efficient engines under the hood.

- **Hot Rods and Rat Rods:** Hot rods, with their stripped-down bodies and roaring engines, remain iconic symbols of Route 66. Rat rods, with their deliberately rough, unfinished look, add an edgy, rebellious vibe to the highway.

- **Modern Classics:** Cars like the Dodge Challenger and Ford Bronco have been reimagined for the 21st century, offering modern drivers the chance to experience a slice of the past with cutting-edge technology and performance.

The Road Trip Experience Today

The vehicles redefining Route 66 are only part of the story—how today's drivers interact with the highway is just as transformative.

- **Technology on the Road:** GPS apps, Route 66 guides, and social media platforms allow drivers to explore the highway with unprecedented ease. Travelers can plan stops at iconic landmarks, navigate detours, and share their journeys in real-time, connecting with a global community of road trip enthusiasts.

- **Blending Adventure and Comfort:** Modern cars, whether electric or restored classics, provide a level of comfort that enhances the Route 66 experience. Air-conditioned cabins, Bluetooth sound systems, and advanced suspension systems make even the longest stretches of the road enjoyable.

- **The Spirit of Exploration:** Despite the technological advancements, the essence of Route 66 remains the same. Drivers today still seek the freedom of the open road, the joy of unexpected discoveries, and the thrill of connecting with the landscapes and communities along the way.

The Future of Route 66 Travel

As automotive technology continues to evolve, the cars on Route 66 will undoubtedly change, but the road's appeal will remain timeless. Autonomous vehicles may one day navigate the twists and turns of the Mother Road, while advances in battery technology will make electric road trips even more accessible. What won't change is the sense of wonder and connection that Route 66 inspires.

- **A Highway for All Generations:** Whether behind the wheel of a 1957 Chevy or a cutting-edge electric roadster, Route 66 offers something for everyone. It's a place where nostalgia meets innovation, and where the journey is always greater than the destination.

Conclusion: A Highway for the Ages

Route 66 has always been more than a road—it's a reflection of the times, adapting and evolving while remaining true to its spirit. The cars redefining Route 66 today, whether restored classics or futuristic EVs, carry forward the legacy of freedom, exploration, and individuality that has made the Mother Road a cultural icon.

As travelers from around the world continue to journey along Route 66, they write new chapters in its story. Behind the wheel of these modern classics, they ensure that the road remains as vibrant and relevant today as it was in its golden age.

Chapter 14: The Future of the Mother Road

A Road at a Crossroads

Route 66 has always been a symbol of progress, freedom, and the open road, but the world of transportation is evolving rapidly. The rise of electric vehicles (EVs), autonomous cars, and smart infrastructure presents both challenges and opportunities for the Mother Road. As technology propels us into the future, Route 66 must navigate the delicate balance between preserving its historic charm and embracing modernization.

This chapter explores what lies ahead for Route 66 in the age of advanced transportation, focusing on how the road can remain relevant while staying true to its legacy.

The Rise of Autonomous Vehicles

Autonomous vehicles (AVs) are no longer a vision of the distant future—they're becoming a reality. As self-driving technology develops, it's poised to revolutionize road travel, including the way we experience Route 66.

- **How AVs Could Transform the Route 66 Experience:**
  Autonomous vehicles promise unparalleled convenience, allowing passengers to enjoy the journey without focusing on driving. Imagine cruising down Route 66 while admiring the landscapes, reading about its history, or enjoying a meal—all without taking your hands off the wheel.

- **The Challenge of Connection:** While AVs offer convenience, some worry they may diminish the personal connection to the road. Driving Route 66 has always been as much about the journey as the destination—will the thrill of navigating its twists and turns be lost when the car does it for you?

- **Adapting the Infrastructure:** For Route 66 to accommodate AVs, its infrastructure may need to be updated. This includes smart road technology, such as sensors and vehicle-to-road communication systems. These changes could modernize the road but might also raise concerns about altering its historic character.

The Electrification of the Road

Electric vehicles are already reshaping the automotive landscape, and their growth has significant implications for Route 66.

- **Charging the Mother Road:** EV drivers rely on charging infrastructure, and Route 66 is responding. Many historic stops, like motels and diners, have begun installing EV charging stations to cater to modern travelers. These stations often blend into the aesthetic of the road, ensuring they don't detract from its nostalgic charm.

- **Sustainability and Revival:** The shift to EVs aligns with a growing emphasis on sustainable tourism. By embracing clean energy, Route 66 can position itself as a leader in environmentally conscious travel, attracting a new generation of eco-conscious road trippers.

- **Challenges of Electrification:** Ensuring that rural and remote stretches of Route 66 are equipped with charging stations is a logistical challenge. Preservationists must also consider how to modernize without undermining the road's historic identity.

Balancing Preservation with Progress

As technology reshapes travel, Route 66 must find a way to evolve without losing its soul.

- **Preserving the Charm:** The neon signs, retro motels, and quirky roadside attractions are integral to Route 66's identity. Preservation efforts must ensure that these landmarks are protected, even as new technologies are integrated into the road's infrastructure.

- **Modernizing Thoughtfully:** Modernization doesn't have to mean losing the past. Smart planning can incorporate technological advancements in ways that respect and enhance Route 66's history. For example, EV charging stations disguised as vintage gas pumps or autonomous vehicle routes that encourage stops at historic sites.

- **Community Involvement:** Local communities along Route 66 should play a central role in its modernization. By involving residents and small business owners in planning efforts, the road can evolve in ways that benefit those who depend on its legacy.

The Role of Tourism in the Future

Tourism will remain a cornerstone of Route 66's identity, but how travelers experience the road is likely to change.

- **Virtual and Augmented Reality Experiences:** Travelers may soon use augmented reality (AR) apps to enhance their journey, overlaying historical images and stories onto the landscapes and landmarks they visit. Imagine pointing your phone at a stretch of road and seeing how it looked in the 1940s, complete with stories of the people who traveled it.

- **Themed EV and AV Road Trips:** Companies might offer specialized Route 66 tours in autonomous or electric vehicles, combining the convenience of technology with curated itineraries that highlight the road's most iconic stops.

- **Global Appeal:** Route 66 will continue to attract international tourists who view it as a quintessential American experience. By modernizing while preserving its charm, the road can remain a global destination for generations to come.

A Vision for the Future

Route 66's future lies in its ability to bridge past and present. As autonomous and electric vehicles become more common, the road can serve as a model for how historic landmarks adapt to modern times without losing their essence.

- **Route 66 as a Testbed:** With its mix of urban, rural, and remote areas, Route 66 is an ideal testing ground for smart infrastructure and sustainable travel initiatives. Piloting such programs along the Mother Road could set an example for other historic highways worldwide.

- **Cultural Resilience:** At its core, Route 66 is about more than technology or infrastructure—it's about the stories, communities, and adventures that define the American road trip. These elements must remain at the heart of its evolution.

Conclusion: A Road Worth Traveling, Always

Route 66 has survived decommissioning, competition from interstates, and the passage of time. Its legacy endures because of the people who cherish it, the businesses that thrive along it, and the travelers who seek its magic.

In the age of autonomous vehicles and electrification, Route 66 has the opportunity to not just survive but to thrive as a symbol of both history and innovation. By balancing preservation with modernization, the Mother Road can remain a destination where the past meets the future—and where every mile continues to inspire.

Chapter 15: Road Trip Guide: Your Journey Down Route 66

Planning the Adventure of a Lifetime

Traveling Route 66 is more than just a road trip; it's a journey through history, culture, and Americana. Whether you're driving the entire 2,448 miles from Chicago to Santa Monica or exploring a single stretch, the Mother Road offers endless opportunities for discovery and adventure. Planning your journey can make the experience even more rewarding, ensuring you don't miss the landmarks, stories, and small-town charm that define this legendary highway.

This chapter serves as your practical guide to navigating Route 66, highlighting must-visit stops, offering tips for a smooth trip, and providing recommendations to make your adventure unforgettable.

Preparing for the Journey

Before hitting the road, a little preparation goes a long way.

- **Choosing Your Route:** While Route 66 has been decommissioned as a federal highway, its path is still navigable through a combination of maps, apps, and signage. Consider whether you'll follow the traditional route, explore detours, or focus on specific sections.

- **The Right Vehicle:** Route 66 accommodates all types of travelers, from those in classic cars to modern EVs and motorcycles.

Choose a vehicle that suits your style and comfort, and make sure it's road-trip-ready with a pre-trip inspection.

- **Packing   Essentials:** Bring a mix of practical and nostalgic items, such as a physical map of Route 66, plenty of water, snacks, and a camera for capturing memories. Don't forget a playlist of Route 66-inspired tunes to set the mood.

Must-Visit Stops Along Route 66

Route 66 is packed with iconic landmarks and hidden gems. Here are some highlights to include on your itinerary:

- **Illinois:**
    - THE STARTING POINT IN CHICAGO: Snap a photo at the "Route 66 Begins" sign near Adams Street.
    - GEMINI GIANT (WILMINGTON): A towering fiberglass figure that's an iconic piece of roadside Americana.
    - ARISTON CAFÉ (LITCHFIELD): One of the oldest restaurants on Route 66, offering a slice of history with your meal.

- **Missouri:**
    - CHAIN OF ROCKS BRIDGE (ST. LOUIS): A historic bridge with a unique 30-degree bend, perfect for a scenic walk.

- o   MERAMEC CAVERNS (STANTON): Explore stunning limestone caves and learn about their history as a hideout for Jesse James.

- **Kansas:**

  - o   GALENA'S HISTORIC DOWNTOWN: Visit Cars on the Route, a service station that inspired Pixar's CARS.

- **Oklahoma:**

  - o   THE ROUND BARN (ARCADIA): A beautifully restored barn that's a Route 66 architectural gem.

  - o   BLUE WHALE OF CATOOSA: A quirky and cheerful roadside attraction perfect for families.

- **Texas:**

  - o   CADILLAC RANCH (AMARILLO): Spray paint your mark on this iconic art installation of buried Cadillacs.

  - o   THE BIG TEXAN STEAK RANCH (AMARILLO): Famous for its 72-ounce steak challenge.

- **New Mexico:**

  - o   TEE PEE CURIOS (TUCUMCARI): A vintage gift shop housed in a quirky teepee-shaped building.

  - o   SANTA FE DETOUR: Explore the rich culture and history of this artistic city.

- **Arizona:**

    - PETRIFIED FOREST NATIONAL PARK: Marvel at ancient fossilized wood and the colorful Painted Desert.

    - WIGWAM MOTEL (HOLBROOK): Stay the night in a teepee-shaped room at this historic motel.

- **California:**

    - ELMER'S BOTTLE TREE RANCH (ORO GRANDE): A whimsical forest of bottle sculptures.

    - SANTA MONICA PIER: The official end of Route 66, where you can celebrate your journey with ocean views.

Tips for a Memorable Journey

To make the most of your Route 66 adventure, keep these tips in mind:

- **Plan for Flexibility:** While it's good to have an itinerary, leave room for spontaneous stops and detours. Some of the best moments happen when you least expect them.

- **Talk to Locals:** Local residents often have the best recommendations for hidden gems, great food, and unique experiences. Don't be afraid to strike up a conversation.

- **Savor the Experience:** Route 66 isn't about racing to the finish line. Take your time, soak in the landscapes, and enjoy the journey.

- **Support Small Businesses:** From family-owned diners to vintage motels, small businesses are the lifeblood of Route 66. Supporting them helps preserve the road's character and history.

- **Capture Memories:** Document your trip with photos, journal entries, or social media posts. These mementos will help you relive the adventure for years to come.

Modern Tools for Navigating Route 66

Technology has made exploring Route 66 easier than ever.

- **Apps and Maps:** Download Route 66-specific apps like Route 66 Navigation, which provide turn-by-turn directions, historical insights, and tips for must-see stops.

- **Online Communities:** Join forums and social media groups dedicated to Route 66 to connect with fellow travelers and gather recommendations.

- **Interactive Guides:** Websites like Historic66.com offer detailed maps, photos, and travel tips to help plan your trip.

The Magic of Route 66 Today

While much of Route 66's charm lies in its history, it's a living road that continues to evolve. Modern travelers bring new energy to the highway, finding joy in both its timeless landmarks and contemporary attractions.

Whether you're drawn to the nostalgia of neon signs and vintage diners or the thrill of discovering hidden gems, Route 66 offers an unparalleled road trip experience. With a little planning and an adventurous spirit, your journey down the Mother Road will be one to remember.

Conclusion: Your Adventure Awaits

Route 66 is more than a road—it's a journey into the heart of America, filled with stories, landmarks, and unforgettable experiences. Whether you're a seasoned road tripper or embarking on your first adventure, this guide will help you make the most of your time on the Mother Road.

So grab your map, fuel up your car, and get ready to create your own story along Route 66. The road is waiting, and the adventure of a lifetime is just a turn away.

Chapter 16: Epilogue: Why Route 66 Endures

The Road That Became a Legend

Route 66 is more than just a stretch of asphalt winding through America's heartland—it's an enduring symbol of freedom, resilience, and the pursuit of adventure. Over nearly a century, the Mother Road has transcended its original purpose as a transportation route, evolving into a cultural icon that resonates far beyond its physical boundaries. It is a road that connects generations, inspiring stories, songs, and dreams.

But why does Route 66 endure? In a world of high-speed interstates and instant gratification, what is it about this historic highway that continues to captivate travelers from across the globe? The answer lies in its unique blend of history, humanity, and the promise of discovery.

A Road Built on Dreams

Route 66 was born in 1926, a time when America was still finding its identity as a nation of mobility and innovation. It connected rural communities to bustling cities, offering opportunities to those who dared to venture westward.

During the Great Depression, it became a lifeline for families fleeing the Dust Bowl, a road to survival and hope. In the post-war boom of the 1950s, it transformed into a playground for families, a route lined with neon signs, quirky motels, and roadside attractions. Through every era, Route 66 has reflected the spirit of the times, adapting to the needs and dreams of those who traveled it.

The Stories That Keep It Alive

What sets Route 66 apart from other highways is the wealth of stories it carries. Every town, every diner, every motel has its own tale to tell, and every traveler adds a chapter to the road's legacy.

- **The Human Connection:** Route 66 thrives on its ability to connect people—not just to the road itself but to one another. From the business owners who pour their hearts into preserving its landmarks to the travelers who share meals and stories along the way, the highway fosters a sense of community that transcends geography.

- **A Canvas for Creativity:** Route 66 has inspired countless works of art, from John Steinbeck's THE GRAPES OF WRATH to Nat King Cole's iconic song. These creative expressions immortalize the road, ensuring that its legend continues to grow with each passing generation.

A Journey, Not a Destination

In today's fast-paced world, Route 66 offers a different kind of travel experience—one that values the journey as much as the destination. The road invites travelers to slow down, savor the landscapes, and embrace the unexpected detours that make life memorable.

- **Nostalgia and Simplicity:**
  Route 66 harkens back to a time when travel was about exploration and connection, not just getting from point A to point B. It offers a nostalgic escape from the pressures of modern life, reminding us of the joy in simple pleasures like a slice of pie at a roadside diner or a chat with a stranger at a gas station.

- **Rediscovery in Every Era:**
  Each generation finds its own reasons to rediscover Route 66. For some, it's a love of classic cars and Americana; for others, it's the thrill of tracing a historic path. Whatever the motivation, the road continues to draw people in, offering a sense of continuity in an ever-changing world.

A Cultural Icon with Global Reach

Route 66's influence extends far beyond the borders of the United States. It has become a symbol of the open road, freedom, and the American dream for people around the world. Travelers from Europe, Asia, and beyond flock to the Mother Road to experience a piece of Americana and immerse themselves in its stories and culture.

- **A Shared Heritage:**
  For international visitors, Route 66 represents a shared heritage of exploration and innovation. It's a chance to experience a piece of history that speaks to universal themes of resilience and the pursuit of opportunity.

- **A Road That Unites:**
  In a time when divisions often dominate headlines, Route 66 serves as a reminder of what unites us—the

love of adventure, the desire to connect, and the enduring hope for a brighter future.

Preservation and Legacy

The efforts to preserve Route 66 underscore its importance as a cultural and historical treasure. Grassroots organizations, local communities, and enthusiasts around the world work tirelessly to protect its landmarks, stories, and spirit.

- **A Living Monument:** Route 66 is not a static relic of the past—it's a living, breathing monument to the American journey. It evolves with the times while honoring its roots, offering travelers a bridge between the past and the future.

Why It Endures

So why does Route 66 endure? Perhaps it's because the road embodies something timeless and universal. It represents the hope of a better tomorrow, the thrill of discovery, and the resilience to keep moving forward even when the road gets tough.

Route 66 endures because it is more than a road—it is a story. It is the story of a nation, of its people, and of the countless journeys taken along its path. And as long as there are dreamers who seek adventure and connection, the Mother Road will continue to inspire.

Final Thoughts: The Road Ahead

As you reflect on the legacy of Route 66, remember that its story is still being written. Each traveler who ventures down its path adds a new layer to its history, ensuring that the road remains vibrant and alive for future generations.

Whether you've experienced Route 66 firsthand or dream of traveling it someday, its spirit invites you to embrace the journey, celebrate the past, and look forward to the adventures that lie ahead. The Mother Road endures because it lives in all of us—a reminder that the road to discovery is never truly finished.

So here's to Route 66: a road that began as a practical necessity and became an enduring legend. The journey continues.

Chapter 17: Appendix: Resources and Further Reading

Discovering More About Route 66

The journey along Route 66 doesn't have to end when you leave the road. Countless resources are available to help you dive deeper into the history, culture, and stories of the Mother Road. From captivating books and documentaries to websites and preservation groups, these resources offer insights into the road's past and its enduring significance.

This appendix provides a curated list of books, films, websites, and organizations that will enrich your understanding of Route 66 and inspire future explorations.

Books About Route 66

Whether you're interested in history, travel, or personal stories, these books capture the essence of the Mother Road:

- **"Route 66: The Mother Road" by Michael Wallis**
  A definitive classic, this book weaves together the history, legends, and enduring appeal of Route 66.

- **"Traveling Route 66" by Nick Freeth**
  A guidebook and photographic journey that highlights the landmarks and stories of Route 66.

- **"Images of 66: A Photographer's Journey" by David Wickline**

This stunning visual exploration captures the spirit of the road through vivid photographs.

- **"Route 66 Adventure Handbook" by Drew Knowles**
  A practical guide for travelers with details about unique attractions, hidden gems, and tips for exploring the highway.

- **"The Grapes of Wrath" by John Steinbeck**
  While not solely about Route 66, this literary masterpiece immortalized the highway as the "Mother Road" during the Dust Bowl era.

Documentaries and Films About Route 66

These documentaries and films provide visual and narrative insight into Route 66's legacy and cultural impact:

- **"Route 66: The Ultimate Road Trip" (PBS)**
  A comprehensive documentary exploring the history, attractions, and significance of Route 66.

- **"America's Mother Road: Route 66" (Smithsonian Channel)**
  This engaging documentary highlights the highway's iconic stops and its role in shaping American culture.

- **"Cars" (Pixar, 2006)**
  An animated film that pays tribute to Route 66's charm and its impact on small-town America.

- **"The Straight Story" (1999)**
  A poignant road trip movie that, while not entirely focused on Route 66, captures the spirit of cross-country travel and self-discovery.

- **"Easy Rider" (1969)**
  This counterculture classic features Route 66 in its exploration of freedom and rebellion.

Websites and Online Resources

For planning trips, learning history, or joining Route 66 communities, these websites are invaluable:

- **Historic66.com**
  A comprehensive guide to Route 66, offering maps, itineraries, and historical information.

- **National Park Service: Route 66 Corridor Preservation Program**
  www.nps.gov/rt66
  Learn about federal efforts to preserve Route 66 landmarks and access resources for historic sites.

- **Route66News.com**
  A news site dedicated to Route 66, featuring updates on preservation efforts, events, and travel tips.

- **RoadsideAmerica.com**
  A fun resource for finding quirky attractions and hidden gems along Route 66.

- **Social Media Groups:**
  - FACEBOOK: Search for Route 66 fan pages and community groups where enthusiasts share stories and recommendations.

- INSTAGRAM: Follow hashtags like #Route66 and #MotherRoad for inspiration and travel photos.

Museums and Visitor Centers

For an immersive Route 66 experience, visit these museums and centers dedicated to preserving its legacy:

- **Route 66 Hall of Fame and Museum (Pontiac, Illinois):**
  Celebrate the history of Route 66 with exhibits on its cultural and historical significance.

- **National Route 66 Museum (Elk City, Oklahoma):**
  A multi-building museum complex showcasing the road's history through vintage vehicles, photos, and artifacts.

- **Route 66 Museum (Kingman, Arizona):**
  Dive into Route 66's role in shaping the West, with interactive exhibits and detailed displays.

- **Oklahoma Route 66 Museum (Clinton, Oklahoma):**
  Focused on the people and stories of the Mother Road, this museum provides an in-depth look at its impact on American life.

- **Old Trails Museum (Winslow, Arizona):**
  A smaller museum highlighting the local history of Route 66 and its connection to the surrounding area.

Preservation Groups and Organizations

These organizations work tirelessly to protect and promote Route 66:

- **Historic Route 66 Associations:** Nearly every state along the route has its own association. These groups organize events, restore landmarks, and advocate for the preservation of the road. Examples include:

  - HISTORIC ROUTE 66 ASSOCIATION OF ARIZONA

  - ILLINOIS ROUTE 66 SCENIC BYWAY

  - NEW MEXICO ROUTE 66 ASSOCIATION

- **Route 66 Alliance:** A national organization that supports preservation and education initiatives along Route 66.

- **National Trust for Historic Preservation:** This broader preservation organization frequently collaborates on Route 66 projects.

- **Contact Information for Preservation Efforts:** Most associations and museums maintain websites or social media pages with detailed contact information for inquiries and donations.

Exploring the Legacy

The resources in this appendix are just the beginning. Whether you're planning your next trip, diving into the history of the Mother Road, or looking for ways to support its preservation, these books, films, websites, and organizations provide a wealth of knowledge and inspiration.

Route 66 endures not just because of its history but because of the people who keep its story alive. By engaging with these resources, you become part of the legacy—ensuring that the Mother Road continues to inspire adventurers for generations to come.

Chapter 18: Photo Gallery

"Route 66 Begins" sign in Chicago

"End of the Trail" sign at Santa Monica Pier

The Chain of Rocks Bridge (St. Louis, Missouri)

Cadillac Ranch (Amarillo, Texas)

The Blue Whale of Catoosa (Oklahoma)

Wigwam Motel (Holbrook, Arizona)

Santa Monica Pier (California)

Blue Swallow Motel (Tucumcari, New Mexico)

Munger Moss Motel (Lebanon, Missouri)

The World's Largest Rocking Chair (Fanning, Missouri)

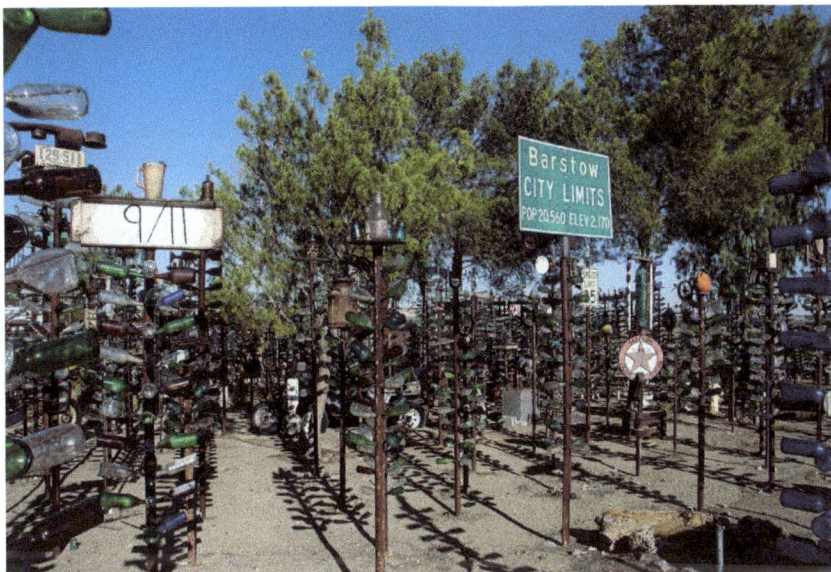

Elmer's Bottle Tree Ranch (Oro Grande, California)

1956 Chevrolet Bel Air

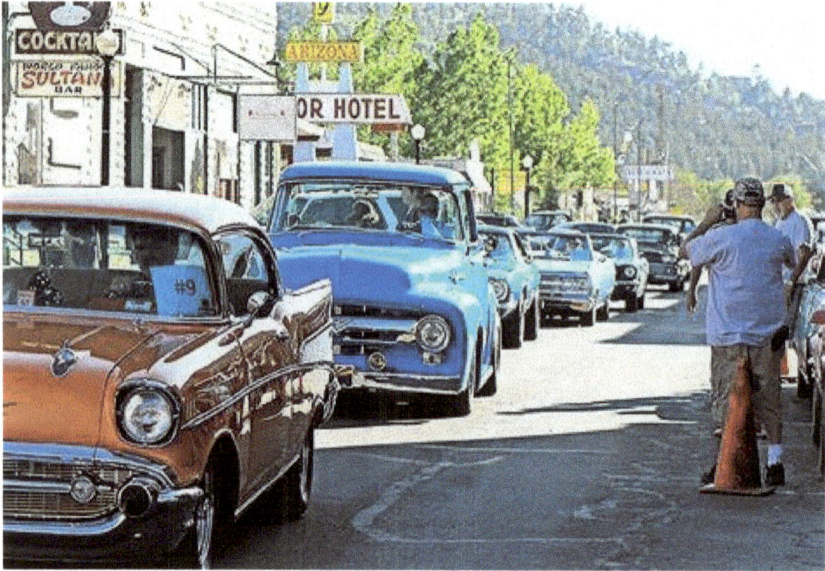

Vintage cars lined up at a Route 66 car show

EV Charging along Route 66

Motorcycles cruising down Route 66

The Painted Desert and Petrified Forest in Arizona

Red Rock State Park in Oklahoma

Endless cornfields in Illinois or rolling plains in Kansas

About the Author

Etienne Psaila, an accomplished author with over two decades of experience, has mastered the art of weaving words across various genres. His journey in the literary world has been marked by a diverse array of publications, demonstrating not only his versatility but also his deep understanding of different thematic landscapes. However, it's in the realm of automotive literature that Etienne truly combines his passions, seamlessly blending his enthusiasm for cars with his innate storytelling abilities.

Specializing in automotive and motorcycle books, Etienne brings to life the world of automobiles through his eloquent prose and an array of stunning, high-quality color photographs. His works are a tribute to the industry, capturing its evolution, technological advancements, and the sheer beauty of vehicles in a manner that is both informative and visually captivating.

A proud alumnus of the University of Malta, Etienne's academic background lays a solid foundation for his meticulous research and factual accuracy. His education has not only enriched his writing but has also fueled his career as a dedicated teacher. In the classroom, just as in his writing, Etienne strives to inspire, inform, and ignite a passion for learning.

As a teacher, Etienne harnesses his experience in writing to engage and educate, bringing the same level of dedication and excellence to his students as he does to his readers. His dual role as an educator and author makes him uniquely positioned to understand and convey complex concepts with clarity and ease, whether in the classroom or through the pages of his books.

Through his literary works, Etienne Psaila continues to leave an indelible mark on the world of automotive literature, captivating car enthusiasts and readers alike with his insightful perspectives and compelling narratives.

Visit www.etiennepsaila.com for more.

9 781763 807419